LET THEM SEE
A WARNING ABOUT YOUR CHILD'S VISION

I0463451

BY
KIMBERLY MCGATH

Vision is a gift most of us take for granted. New parents have so many responsibilities and often are inundated with the dos and don'ts of parenting, so understandably eye checkups can fall to the back of the "to-do" list.

This journey will leave you with an indelible impression — your child's vision is not a guarantee.

This thought provoking account will "open your eyes" to the importance of making your child's eye health a top priority, and will give you the tools to do so.

A portion of the proceeds of this book will be donated to Prevent Blindness, www.preventblindness.org.

LET THEM SEE
Kimberly McGath
Copyright 2016

ISBN: 978-1-365-32500-7
ISBN eBook: 978-1-365-32526-7

Lulu Publishing Services, First Edition 08/12/16

DEDICATION

This book is dedicated to all blind persons.

ACKNOWLEDGEMENTS

To my dearest author friends Sue Coletta, a.k.a. Daphners, and Shaggs — you know who you are. Thanks for all of the support, encouragement, inspiration, and the endless laughter. You have touched my soul, leaving an imprint on my heart, and as Sue would say… "I love you both to the moon and back."

To Author D. Allen Rutherford — thanks for all the advice and your friendship, which I cherish. Your generous spirit shines and I am so grateful for all your advice.

Special thanks to my dear friends Ellen Cranos, a.k.a. Beebo, Stephanie Angel, Patricia Yusefzadeh, Sarah Reeser, Colleen Paice, Sophie Pembleton, and author Zoe Wright for always having my back. I am so blessed to have you in my life.

All of my sincere appreciation to an exceptional crew — Doctor Patricia Blanco, Doctor Janet Davis, Doctor Arysol Soltero, Dr. Thomas Schwartz, and Doctor James Banta. Thanks so much for your expertise and for putting up with me!

All my gratitude to my family for your unconditional love. You are my everything!

And of course, thanks to God for giving me the courage to write this, and for *seeing* our family through some dark times; all my love and never-ending gratitude.

FOREWORD

At first, I agonized over whether or not I should write this book. After all, this type of material is a drastic departure from true crime, and reliving some of these events is sure to stir up some painful memories; but I felt it would be selfish of me not to share my journey. After all, people tend to learn more from others' mistakes and experiences, so I felt it was important to make the effort.

Silence kills. If there is even a slight chance the information provided could prevent one child from enduring the devastating consequences of eye disease or spare one parent the anguish and helplessness in caring for a suffering child, then the fruits of this labor will be realized.

Any statements made in this book regarding medical errors should not be construed as criticisms of the medical profession or any of its members. It was important for me to be honest, and describe the events as they occurred. Parents and doctors alike are flawed human beings and we can all learn and strive to do better. Throughout my journey, I have met some exceptional doctors who have my utmost admiration and respect.

Every effort was made to write this simply and avoid medical jargon as much as possible. My goal was for this work to be a "guide" and not a literary piece or medical source. The mission here is just to create awareness and prevent blindness.

In answer to some of my readers, my crime novel is in the works, and will be written in a drastically different voice than this guide. For more updates and information, please visit — kimberlymcgath.com.

Prologue

Close your eyes. Now imagine. You can never open them again. The sun will never rise, the night will never cease, the sun will never come out, no, not even tomorrow.

For some, there is no escaping this eternal darkness; a life sentence for a crime not committed.

Somewhere out there is a mother being told her child may go blind. How I wish I could hold her, tell her everything will be okay, share with her what I've learned, so she doesn't make the same mistakes. Wanting so much to tell her "our" successes too, so she and her child could benefit from them, but I cannot. Sadly, she is unknown to me, and this is the only way I can reach her...and anyone else for that matter.

It is my hope, that by spreading the word, together, we can make a difference. While not in all, prevention is key in many instances. If one case is detected early, and a child is spared from blindness, well, this is a reason unlike any other to celebrate.

Now it's time, to open your eyes, and Let Them See.

Kimberly McGath

Chapter 1 – A horse of a different color

Exposed — like being randomly selected for a strip search at an airport by a sweaty official — that's the all-encompassing feeling that accompanies me going back to that dark era. But just as James Kirk had to go back in time to find an extinct whale, I too have to scour the murky recesses of my mind, something I have been unwilling to do until now. Knowing you'll be with me is comforting. At least, I don't have to enter the abyss alone.

Back in those days, I was a stay-at-home mom, a drastic change from working in the dynamic world of law enforcement. My biggest concerns were the selection of educational programs and processing organic baby food, but having three kids in four years was no easy task, even when things were running smoothly.

I'll never forget that day. Like a movie playing in my brain, I can see the waiting room in the panorama. "Augustine," the nurse called out. Finally! We had been waiting for what seemed like forever, and trying to entertain three toddlers was a challenge. Oh, how the hours dragged.

My daughter Lee, and son Travis ages five and three at the time, crammed into one seat with a Dr. Seuss book, while I sat in the examination chair. The words still echo in her innocent voice — *Left foot, left foot, right foot, right. Feet in the morning, feet at night.* A haunting rhyme that transports me back to that place...that time.

With my two-year-old in my lap, the eye doctor pushed the phoropter towards him. My first instinct was to smack it away. The medieval-looking instrument's approach reminded me of Darth Vader's mask descending on young Skywalker's head. Thinking he'd squirm or freak-out, I held on tight, but he just clung to his blankie and sucked his thumb.

Stern but courteous, the doc looked professional, almost angelic, in his white coat. As if that wasn't enough to ease my mind, he was a friend of my father's. My dad was a very popular pediatrician in my hometown for thirty-two years, so I knew we were in good hands. There was no reason to doubt this M.D. with his impressive credentials displayed about the office.

"Pink-eye," he said as he spun around in his medical stool.

"Are you sure?" I asked. "He's had this for over a week now."

Cocking his chin down and crinkling his brow, he gave me that look as if to say: *You're kidding right? Relax, silly woman.* As if hypnotically, his assuring words convinced me, at least briefly, there was no cause for concern. So, like most moms, I followed the doctor's instructions and put the matter aside.

An uneventful week passed, and a sick feeling snuck into my being like I had missed something. It's likely most parents can relate; that moment when you realize something's been in front of you the whole time, but you just got too busy to notice or...denial maybe? Either way, my gut was telling me something was really wrong, and my intellect was abuzz. My kids had contracted "pink-eye" before, which usually spread through the house like wildfire. But this time...only my youngest, Augustine, had it.

Frustrated, I brought Augi back to the same ophthalmologist, and he didn't seem too concerned. "Allergies," he said whipping out his prescription pad. "Here's another ophthalmic solution."

Admittedly, I was skeptical at first, but then I felt relieved. *Oh allergies, phew.* Of course, the doctor knew what was best, and maybe Augi did have Hay Fever. How did I miss that? So, again, I followed the orders and tried to put the matter out of my mind.

Another week went by and my baby's eyes still looked bloodshot. That sinking visceral feeling returned, reaching a new apex. Something was amiss. For whatever reason, I knew I had to change course. Even though my father always warned me not to, I consulted the internet. It didn't take me long to find the dreaded answer. Panic set in.

Of course, it was the weekend, as it always seemed to be when there was an urgent medical need. Placing the call to the doc's service at least subsided my anxiety for a few moments. It felt good to take some action. An hour crept by. Time seemed to pass differently waiting for the doctor to ring, like a ripple in the space-time continuum. When the phone finally chimed, I bypassed the formalities and asked one question: "Do you think this could be red-eye?"

After a few seconds of silence, his response and tone scared me. "Bring him in... first thing tomorrow morning."

That was a long, harrowing night. The dam had broken. Frightening questions flooded my mind, and there were no answers to accompany them. What was wrong with his eyes? Are they damaged? What caused this? Was it my fault?

Mounting like never before, my anxiety surged. Wishing I had trusted those intuitive tugs from the onset, it was terrifying to know precious time had passed. It actually would have been better to have done nothing. Not only was he receiving unnecessary medications that aggravated his eyes, he was not being given the crucial treatment he needed to slow or halt the intruders in his eyes.

Something sinister had been hurting my child and on my watch. It was as if he had been falling under water, and no one went in for the rescue. Self-doubt consumed me, and the mommy-guilt took root like cancer.

The morning couldn't come fast enough. Compliant, I brought Augi in to see this same physician, *again*, and he confirmed my diagnosis was correct. Seething over his incompetence, I imagined yelling at him, creating some new swear words in the process.

The shock was overwhelming. Feeling betrayed and woozy, I rushed out to the parking lot and immediately called my dad.

Even though the sound of his voice was comforting, what he told me was not. "I've never seen *that* in my practice...not in all these years."

The diagnosis is a bad word in my house and one that is difficult to write, let alone speak. But I promise you... I will tell you...soon.

As it turns out, the difference between "pink eye" and "red eye" is rather dramatic. While to a parent or even a doctor, the conditions may appear similar, the cause, prognosis, and treatment are drastically different.

With the exception of the classic yellow crust and ooze associated with viral or bacterial conjunctivitis, both conditions make the eyes appear "red" or "pink."

How I wished the doc's initial diagnosis was correct. This subtle difference in "hue" so to speak would drastically change the course of our lives. My Augi was a horse of a different color.

Chapter 2 – Can you hear the hoofbeats?

The shocking news shattered me, eclipsed my soul, darkened my way of thinking. It was more than just the terrifying feeling that something was wrong with my child — it was also the anger over the doctor's handling of his condition, leading to the delay in diagnosis. My head was spinning. It was as if everything I had ever been taught about doctors was wrong — a flawed indoctrination. Physicians were Gods, weren't they? There was never a reason to question the men behind the curtain; they took an oath, after all. At least I had never had a strong reason to doubt a "medicine man" before, so the thought had never occurred with any substance.

Without trying to be too dramatic, it was sort of a crisis of faith, but not in God, in medicine. Feeling betrayed, like I had just found out a lover had cheated on me, was a nasty sentiment I couldn't shake. Only this was worse. This wasn't just my feelings being hurt; my child's vision was at stake.

Anger consumed me as if it infected my very bones. My child had been on the wrong drops for over a month and even worse, I had no one to lean on for guidance. There was no one to turn to...no one to trust. I couldn't find anyone who had ever even heard of that ugly word, yet alone had any experience with it to guide me.

This was a first for me. When I became pregnant for the first time, I had plenty of friends and family to give me the scoop, not to mention the countless books on the subject. But this was different and left me feeling more solitary than I had ever felt.

Even though I had been warned not to, I knew I had to investigate, on my own. There would be no resting, not when it came to my child's sight. Thinking back on his young life became a hobby. What had I missed? Augi had experienced some other medical complications before the "red-eye" diagnosis, but I hadn't thought too much of it. He'd had one serious illness that had required hospitalization and a sickening rash followed. Sure I was traumatized by his suffering but didn't consider at that point that something may be *seriously* different about him.

Feeling burdened, my first step in my investigation was simple: obtain all of his medical records. After a few phone calls, some signing of papers, and the shelling out of some cash, I received about two hundred pages, and what I found was nothing shy of astonishing. Augi had been diagnosed with some conditions that no one had ever disclosed to me. To add to the sting, one of them stood out for an obvious reason — the syndrome noted by the pediatrician was known to cause uveitis. In her handwriting, three illuminating words jumped off the page, "**Steven Johnson's syndrome**." Anger was consumed by its fiercer cousin, and fury took over. Why didn't she tell me?

Feeling motivated by my rage, I immediately switched pediatricians and ophthalmologists. Unfortunately, these second doctors were not any better, and actually, the pediatrician was incontrovertibly worse. It didn't take me long to realize that he didn't listen to anything I had to say and well, basically, "blew me off." Augi's course of treatment was not going as it should, as he wasn't improving. My mistakes were mounting and my poor son was paying the price.

My heart was heavy and I was focused on his convalescence, but I was making matters worse, not better. Augi's first pediatrician was really a good doctor, but I was young at the time, and a conflict-avoider. Wishing I had just sat down and communicated my concerns to her was fruitless — there was no going back. Now, I had wasted a year...time I could never get recoup; crucial to his recovery. How could I have screwed things up this badly?

Knowing I had to find another doctor, a good doctor, I did my research. As luck would have it, I found a new one...the best. This doctor, no doubt, on multiple occasions with her skill and expertise, made some lifesaving decisions on my children's' behalf. Even with all the mistakes, I would make throughout the years, the time I spent finding her was well worth it.

At Augi's first appointment, she stunned me when she said, "When doctors hear hoofbeats, we often think of horses, not zebras." What she meant was that my child had something rare and that's why the other doctors did not understand him or his illness.

Feeling comforted and terrified all at the same time, at least I knew she understood. In front of me was a doctor who was FINALLY listening to me, even agreeing with me, but that meant there was something wrong with my child. Feeling hope for the first time, I felt some relief, but this was overshadowed by a gripping feeling — what kind of a zebra was he?

Chapter 3 – Right out of the gate

The horse and jockey that respond well to the starting bell tend to have an advantage at the derby. The same of course is true in medicine. Whether it is cancer or eye disease, such as in Augi's case, early diagnosis and proper treatment are critical for a successful prognosis. Ask any oncologist, and they'll tell you, better to detect cancer at stage I than stage IV.

Initially, Augi was misdiagnosed and the delay of a month, maybe more, combined with a less than adequate treatment protocol, had a devastating impact on the course of his illness. What could have been a few irritating weeks of eye-drops turned into a grueling decade or more of appointments, painful symptoms, and uncertainty. Of course, I didn't know any of this at the time, nor did a lot of his doctors, despite their best efforts.

It is useless and even difficult to ascribe fault, although for years I blamed myself and the doctors. In my defense, no one ever gave me a road map on how to successfully navigate the system or make decisions concerning a child with the rare disease.

As for the doctors, whilst it is their profession, I understand they heard hoofbeats and thought horses. Those who initially examined Augustine naturally thought pink-eye, not red-eye, for the former is much more common. Still, for years I was bitter... particularly over the fact that little Augi had to suffer for our mistakes. Experts later confirmed had he been diagnosed and treated swiftly, his disease would have remised almost immediately.

The good news is — for your child, your grandchild, your niece, or your nephew, these mistakes can be avoided, and I'm going to give you the necessary steps to do so. So, I guess it's time to tell you that awful word.

Chapter 4 – Dressage is not a dirty word

To a horseman or horsewoman in my case, dressage is an ordinary word and in my vernacular. But when I served on mounted patrol and discussed training events with friends or co-workers, I'd often hear, "What's dressage?" It always surprised me few had ever heard of the term. But it actually makes sense. If you're not an equestrian or into horses, why would it come up?

Same is true for *the dirty word*, the one I hate to say or write, but in this instance, seems necessary — uveitis.

James Joyce, one of the most influential novelists, suffered from uveitis — pronounced yoo-vee-**ahy**-tis. Ironically, one of his middle names is Augustine. It seems my son and the avant-garde writer have two things in common. Even whilst becoming well immersed in the medical literature, Joyce is the only famous person I came across that suffered from the disease.

While reviewing our family history, I discovered my uncle had a brief bout in his twenties; but other than that, I never met anyone that had ever been afflicted with the condition. There was no one else in the world I knew personally that suffered from uveitis, other than my son. Even more isolating, I never met anyone who heard of the disease.

If you think about how many times you've heard about someone having cancer or diabetes, for example, it emphasizes the contrast between a common diagnosis or an unusual one. Even if it's not someone you know, chances are, you've seen someone on television reference those well-known conditions.

Not that the topic is amusing, but people often murmur a disease like a mother who whispers "drugs" in *St. Elmo's Fire*. Leukemia is a dirty word like uveitis, but it is a term in which people are accustomed. When I mention my son's condition, the common reaction is, "What's that?"

Cancer is well known, in the "common speak," and therefore fundraising efforts are more successful, and so the public is more informed. In contrast, most people are uninformed about uveitis, resulting in reduced funding, and sadly, less research. If my son had never contracted this revolting illness; it is doubtful I would have ever heard of it either...wished I never had.

So, what is uveitis? Quite simply, uveitis is arthritis of the eyes, specifically the uvea. Just as rheumatoid arthritis damages the joints, uveitis wreaks havoc on the eyes, and can lead to blindness.

Whilst the eyes are technically considered organs, in a way they behave like joints. It makes me wonder if perhaps this is why almost all of the diseases that cause inflammation in the joints also affect the eyes. In fact, the circumduction of the eyes resembles the movements of a ball and socket joint.

One of the reasons most have never heard of uveitis is rather elementary — most people have never heard of the uvea. The uvea is the center of the eye comprised of the iris, (the colored part of the eye), the choroid layer, (which is comprised of blood vessels and connective tissue between the sclera and the retina, the front and the back of the eye respectively), and the ciliary body, (the secreted of transparent liquid). Whilst this may sound complicated, think of the uvea as being behind the sclera, (the white part of the eye), and in front of the retina, (the back of the eye).

Other than physicians and anatomists, most people don't think about the various structures inside of the eye or the ear for that matter. So, naturally, the term uvea is not thrown around in ordinary speech. If someone hears the word hammer, rarely would the ear come to mind.

Doctors use a lot of big words and unfortunately, I will have to as well to explain certain aspects of this disease, but I will try to keep things as rudimentary as possible. For those interested in the particulars or a more in-depth analysis, there are many resources available online. The purpose here is just to give enough information so parents can identify and recognize the condition.

To simplify some of the "lingo," a lot of terms end in itis or osis. The suffix "itis" generally denotes inflammation in an organ or part, while the latter refers to an abnormal state. So uveitis is literally inflammation of the center of the eye.

Before my son's diagnosis, I don't think I ever even considered or imagined the possibility of the center of an eye becoming inflamed. It's actually difficult to imagine now. Indeed, it is still a very abstract concept.

Most people are familiar with the terms arthritis and rheumatoid arthritis. Perhaps an image of deformed knuckles or an elderly person comes to mind, but unfortunately, children suffer from this condition as well. If such progresses and is left untreated, the process can be destructive to the joints. Uveitis is similar, but the damage is to the eyes, organs necessary for sight.

To complicate the understanding of uveitis there are different terms used to describe the condition such as iritis, (inflammation of the iris), or iridocyclitis. Some doctors use these terms interchangeably. Uveitis can also be categorized as anterior, (the middle of the eye) or posterior, (the back of the eye). In fact, there are so many terms associated with this condition, it would be impractical to cover them all, but there are some features of the condition that can assist in determining the cause or nature of the uveitis, and I will cover those later.

As if it was not alarming enough that uveitis itself can cause so many complications to arise such as glaucoma and cataracts, the treatment can be just as damaging. Often steroid drops are prescribed to reduce inflammation caused by uveitis. Ironically, such medication can also cause glaucoma, one of the very dreaded aspects of the condition itself. One of the almost certain side-effects of prolonged steroid use is the formation cataracts.

As you can imagine, cataracts lead to blurry vision. The cloudiness obstructs sight and is often described as looking through foggy or frosted glass. Seeing halos around lights can also occur with this condition. Headaches and photophobia are commonly reported symptoms. Severe cataracts can be visible externally and as such are considered a medical "sign" as they can be observed.

While most consider cataract surgery routine, I cannot stress enough the importance of selecting a doctor or center that specializes in such procedures, particularly for children. If not performed correctly, such surgeries can have devastating consequences resulting in loss of vision.

Fortunately, cataracts can be surgically corrected, although artificial lenses do not identically mimic human lenses. Patients who undergo cataract surgery can opt for various types of lenses, but such do not ever completely replicate normal human vision. Depth perception is reduced, although glasses can be prescribed to minimize such loss.

Steroid drops used to treat uveitis, along with the condition itself, can cause pressure to rise in the eyes, causing glaucoma, a sometimes painful condition depending upon the type. Loss of peripheral vision may be the first sign, and such should be considered a medical emergency. Severe eye pain or sudden blurriness in vision should prompt immediate action.

Sudden vision loss can also be caused by optic neuritis which can be accompanied by pain. Note the suffix of "itis" again. Literally, this term refers to inflammation of the optic nerve. Optic neuritis can occur unilaterally or bilaterally, depending on the cause and presentation. Symptoms include photophobia (literally fear of light, but avoidance of light is more accurate), pain, loss of color vision, particularly red hues, and reduction in night vision. Optic neuritis can occur alone or as a result of chronic uveitis.

Synechia, another condition common with uveitis, can cause the iris (the colored part of the eye) to stick to the cornea, and this, in turn, can also cause glaucoma. The treatment for this condition is to continually dilate the eyes with drops. (Augi's eyes were dilated for years, making learning and reading a challenge.) This process can also cause the pupil to permanently become misshapen.

The inflammation in the eyes caused by uveitis can become so out of control that shots in the eyes are recommended. Such aggressive treatments carry an even higher risk of side effects, and can be quite traumatic, and particularly in small children.

In addition to local treatments, there are also "systemic" treatments aimed at lowering the inflammation in the eyes by treating the body as a "whole." Such treatments, either intentionally or by default, can potentially treat a related condition, such as Juvenile Rheumatoid Arthritis. These treatments include methotrexate (a chemotherapy of sorts) and NSAIDS (non-steroidal anti-inflammatory drugs). The concern here, of course, is these drugs can damage the liver and lower the immune system's ability to fight off infection.

Another systemic treatment often utilized is oral steroids. There are many known side-effects including damage to internal organs, an increased risk of certain metabolic disorders, and ironically, glaucoma. Cushing's syndrome, also known as hypercortisolism, is an endocrinological disturbance in which the adrenal glands overproduce cortisol. Symptoms include central weight gain, hypertension, moon-face, and a lump in the back amongst others. While it is known steroids can induce Cushing's syndrome, it is unknown how much of an affect ophthalmic solutions containing steroids affect the body.

Treatments aimed at managing uveitis can cause just as much, if not more, damage to the patient, and are never guaranteed to be effective. Even worse, if a patient displays signs of uveitis, most likely, there is an underlying systemic condition accompanying or causing the uveitis, and such can be difficult to determine, complicating the course of treatment.

So, when treating uveitis, you're basically damned if you do...damned if you don't. The condition and the treatment both cause similar damage to the eyes — a real medical catch-22.

If you or a loved one has been diagnosed with uveitis or a similar condition, don't fret. Since I've already negotiated "the system," I will give you some great tips on how to navigate this illness. There is hope.

Chapter 5 – Hold your horses!

The most important information that I want to relate to expectant mothers, parents, grandparents, teachers, and the public, is how to recognize eye disease. Of course, I don't want to make anyone feel paranoid or overly concerned but want to create awareness. Knowledge is power!

Most pediatricians recommend parents take their children to an eye doctor by age four. However; in Augi's case, as is with many other children, this can be way too late. So, there are some signs and symptoms to be aware of, in case a problem occurs before this age. Also, I would highly recommend once you are ready to take your child for an eye examination, you see an ophthalmologist, not an optometrist. Optometrists are great when you need glasses or contacts, but in case there is a problem, particularly a rare one, it is imperative to see a doctor. Selecting the "right" doctor is also crucial, and I will elaborate upon such further in a later chapter.

Most parents have seen the warnings on social media regarding retinoblastoma, a rare eye cancer. A photograph can show leukocoria, which shows the affected eye with a white orb, while the unaffected eye is seen with the red-eye appearance or "eyeshine" typically captured with flash photography. This condition can also present itself with strabismus, also known as "cross-eyes." Even strabismus alone without any accompanying disease needs to be detected and treated early. Whilst it is normal for babies to cross their eyes occasionally, this can be a sign that your child needs to be seen. The treatment for strabismus varies including eye patches, glasses, and even surgery to correct the extraocular muscles if the case has been detected late or is severe.

Another sign one can glean from photographs is excessive squinting. Looking back on certain pictures of Augi, I noticed he squinted often, most likely from the flash. Photophobia (fear of light) can accompany eye conditions such as uveitis and should not be taken lightly. While some blinking or squinting from flash photography is normal, excessive behavior should be evaluated.

Headaches, particularly if severe or frequent can also be indicative of eye disease. Some children suffer from chronic headaches, and it is important to determine the cause of such. Loss of vision or inflammation can trigger many different types of headaches, particularly behind the eyes. Even common refractive errors such as hyperopia (farsightedness) and myopia (near-sightedness) can trigger headaches in children.

Any discoloration of the sclera (the whites of the eyes) can signal a problem. Episcleritis (inflammation of the sclera) can cause a bloodshot appearance to the eyes. There are also other conditions, although rare, in which the sclera can take on a bluish hue, indicating a serious systemic condition.

Excessive tearing can also be an indication of a problem. A blocked tear duct is a common occurrence in infants. Pediatricians often recommend applying a warm compress to the affected eye, and such condition usually resolves on its own. In severe cases, a procedure to unblock the duct is performed.

Excessive rubbing of the eyes can also indicate a problem with vision and should prompt a trip to the ophthalmologist. Drooping of the eyelids, ptosis, is also another source of concern, and can signal a neurological cause or sequelae of eye disease. Sudden onset ptosis should be considered a medical emergency as such can be indicative of a stroke or transient ischemic attack.

Bottom line, if you notice any color changes to the eyes, blinking, tearing, photophobia, excessive headaches, rubbing of the eyes, drooping of the lids, eye-crossing, squinting, or other signs or symptoms involving the eyes, it's time to consult an ophthalmologist.

Furthermore, any children with rheumatic conditions such as arthritis or spondyloarthropathy should get their eyes examined biannually for life. It is also vital for parents to know and be able to recognize the symptoms of uveitis so it can be detected and treated early. In conditions such as these, there is a high corollary factor with uveitis, and as such, these children have a greater risk of contracting uveitis. Therefore, it is imperative to keep a watchful eye to protect your child's vision. Once your children reach young adulthood, it is also essential to teach them the warning signs, so they can self-monitor their condition.

As you can see, understanding uveitis and knowing the symptoms is a very important step in preventing blindness in children and adults alike, but this alone is not enough. The doctor you choose for your child is a critical decision and one I will address next.

Chapter 6 – Straight from the horse's mouth!

Finding the right doctor, whether it be a general practitioner, such as a pediatrician, or a specialist, is of the utmost importance. If I had found the right ophthalmologist for Augustine the first time, he could have been spared so much. Of course, I had no way of knowing at the time he had a rare eye disease and such, but if I could go back, this would be the crucial decision I would change. There is no doubt in my mind that if his first doctor caught and treated the uveitis immediately, it would have remised rather quickly.

My point here is that if you're going to eventually take your child or loved one for an eye checkup, why not have the best? It is essentially more prudent to invest the time up front and find the right doctor, preferably one that at least has treated rare conditions. So if something comes up, that doctor can at least recognize a problem. Think of it as an insurance policy. You purchase flood coverage, for example, and spend a lot of money, just in the event something happens. So finding the best possible doctor is a worthwhile investment and will provide you with peace of mind.

There are many qualities to look for when selecting the right doctor for your child. Open-mindedness is essential, particularly when your child is afflicted with an unknown condition. Experience in diagnosing and treating rare conditions is important, as without such familiarity an accurate diagnosis is less likely. A good bedside manner is preferable, especially if you are going to be spending a lot of time conversing with the physician about your child, and how to manage the chronic condition.

Once you have narrowed down your selection, it is a worthwhile investment to research the doctor's rating. There are various websites devoted to such, and usually, patients will write testimonials as well. Depending on the size of your community, it may even be possible to speak with parents or patients in person about their experiences. There are also ways to check a doctor's medical license status, and if there have been any lawsuits filed or complaints lodged against such professional.

One of Augi's eye doctors, even after he was diagnosed with uveitis, missed the inflammatory "cells" for months. As you can see, we had more than one disappointing experience with more than one doctor. This still baffles me, but demonstrates the frightening reality: if doctors have not treated or seen a certain condition, it is more likely than are not qualified to diagnose or treat patients with such affliction. So even if a doctor does not have to make the initial diagnosis, he or she may still not be qualified to treat the illness properly.

As in any profession, there are examples on the extreme ends of the spectrum. There are exceptional doctors who are intelligent, kind, and skilled and there are those who commit malpractice, fraud, and manslaughter. There are those in the middle too who are well-intentioned but lack the necessary skills or education to excel in their field. This is not at all a criticism of doctors; it is just a statement of reality. There are varying degrees of greatness or lack thereof in any field. When it comes to your child or loved one, only the best will do.

Vision is essential, so vital in fact that the selection of the best or "right" eye doctor may be one of the most important decisions you make for your child, so select wisely.

Chapter 7 – Don't look a gift horse in the mouth!

Even if some parents forget their child's first eye checkup, most preschools, and elementary schools offer a vision screening. In fact, I learned this when I received a letter from Augi's preschool stating he had failed his vision test. Even though I already knew he had already suffered a loss of visual acuity due to his disease, it still was upsetting to read.

These screenings, while a good start, typically only identify vision loss. The personnel employed to conduct such screenings are not trained in detecting eye diseases such as uveitis, vitritis, or retinitis, nor are they provided the proper equipment to do so.

Even though the screenings can detect some vision issues, thorough eye examinations are still essential to uncover other conditions. Unfortunately, resources are limited often and in the developing world, scarce. While it may be impractical, advocating for eye checkups for all children certainly is a goal worthy of our best efforts.

Along with screenings and eye checkups, public awareness is also vital in preventing blindness. The only public service announcement I have ever seen regarding the eyes is a commercial regarding diabetes-related blindness. While this is a great cause, the lack of advertisements for eye disease, in general, is not acceptable. An increase in funding is needed to bolster public awareness about the dangers our children face, particularly when the consequences are so dire. With the time spent on social media and our technological know-how, it's inexcusable for us not to try and spread the word.

Educating doctors, parents, and teachers should also be a high priority as those professionals interact with our children the most. Members of these fields are on the "frontline" and can have the most impact by recognizing the signs and symptoms of eye disease.

In law enforcement, we trained and prepared for the worst case scenarios, not the best. Even when the probability was low that such an event would occur, such as an active shooter or a chemical spill, we trained nonetheless — with much money spent and time expended. Doctors should be trained similarly. While I understand it is unrealistic for a doctor to always think the worst, or picture a "zebra" when hearing hoofbeats, it should at least be in the differential. It's better to have more tools in the box, so at least that odd item is needed, even if it's only once.

More courses in recognizing rare diseases should become a part of the mandatory curriculum in medical school. Lectures on diagnosing and treating atypical conditions should also be included in seminars and online training as part of licensure requirements and maintenance. Reminders of uncommon afflictions are necessary because of years, even decades can pass before such a condition is seen in a hospital or office setting. So even if exceptional diseases are taught in medical school, doctors may forget their significance if not seen regularly in practice. Therefore, periodic training in rare disease is imperative.

Teachers are also on the forefront and have a different perspective than parents. Having a brief education or training session on detecting eye disease could identify many cases which otherwise may have gone undetected. This doesn't have to be an extensive understanding of uveitis or other eye diseases.

Just a brief explanation, a simple awareness, could plant a seed in a teacher's mind that could bear much fruit. Teachers spend an exorbitant amount of time with their students, even more so during the elementary years. Teachers who receive awareness training would be more likely to recognize a problem. For example, a teacher may notice a student rubbing their eyes every day. This could trigger a recollection of the warning signs, and the teacher could notify the child's parents. This teacher's intervention could be paramount to that child's future.

A teacher who attends an hour lecture or even reads a pamphlet could learn enough to intervene and spare a child from living in a world of darkness. Teachers receive training in recognizing child abuse, sexual abuse, even learning disabilities, so why not include a basic understanding of recognizing a disease that can deprive students of one of the most necessary tool for learning — sight.

And for you animal lovers like me, uveitis is not just a disease that occurs in humans, but we'll get to that in a later chapter.

Chapter 8 – Horse sense

The best advice I can share from my experience is to trust your gut. If you sense something is wrong — act...do something. Don't let people convince you that you are just worried or over-thinking things. If you truly sense a problem, don't be ashamed to seek a professional opinion. Better to hear all is fine, then to do nothing, and suffer the consequences.

This is true whether you suspect a new condition, or you don't trust a doctor's opinion. The truth is...it is just that — an opinion. It is so important to be your own advocate; even more important to be that for a child who cannot act on his or her own behalf.

Having said that, if you invest the time to find the right doctor, hopefully, you won't need a second opinion. Switching doctors are expensive, time-consuming, and interferes with the ability to develop the ever so important patient-doctor relationship; the deeper the bond, the greater the care.

When I was a just a few years old, I overheard my parents talking. My dad remarked how his CF (cystic fibrosis) parents knew more about the disease than he did. Well, this makes sense and my dad understood. These moms, of course, immersed themselves in cystic fibrosis literature, wanting to help their precious children in every way possible. But my dad, being a general practitioner, had to know a little about all pediatric conditions. With the wealth of knowledge available now, it is impractical, if not impossible, for a doctor to know about every condition, every symptom, every treatment. In fact, there are so many discoveries on a daily basis, it would be impossible for any one doctor to keep abreast of all the research, old and new.

Whether it is uveitis, diabetes, or spondyloarthropathy, chances are you will become an "expert" in your child's condition. Initially, it may be wise to dial back the "lingo," particularly when consulting a new doctor or specialist. Some doctors find it very off-putting when a parent uses medical terminology, and may even dismiss you as a result. Once you establish a rapport with a doctor, you'll be able to gauge their sensitivity on the matter.

You may find that some specialists you consult may not have even heard of your child's condition, particularly if it's a rare one like uveitis.

Shockingly, I have experienced this. and if that happens to you...I suggest you just walk out the door.

Chapter 9 – Horses for courses

If you're in the unfortunate position that I was all those years ago and have just heard that ugly word, uveitis, I'm here for you. It's horrible news but there are many things you can do to increase your child's chances of beating this ugly disease.

As I expressed earlier, the finding of a local eye doctor is critical but finding an eye center is just as important and I'll explain why.

After I found the best pediatrician, she recommended we take Augi to an ophthalmologist at our local children's hospital. This new eye doctor was also a friend of my dad's. It was only about an hour drive and he was a really nice, competent doctor, but after several months, I didn't notice much improvement. Augi's eyes were dilated all day...every day. This just didn't seem right to me. How was he ever going to read?

After going through several doctors and such, I knew I had to get back on the internet. After a few clicks, I discovered there are a few eye centers in the country that specialize in rare eye diseases such as uveitis. As luck would have it, one of them was in Florida, our home state. There are a few others, most notably, Wills Eye Institute in Philadelphia and Massachusetts Eye Research and Surgery Institution in Boston.

After researching the various doctors, I scheduled an appointment with a retinal specialist at the Bascom Palmer Eye Institute.

Twenty questions helped us pass the five-hour drive to Miami. Each member of the family knew all my questions would be about Star Wars, which just made us all laugh when it was my turn. The first question would always be, "Is it a character from Star Wars?"

Our hotel room was cozy and had a pool, which I knew the kids would love. Lee, Travis, and Augi were having such a blast that night, swimming and playing Marco Polo... that is until the crash. Travis swam fast under water but did so with his eyes closed. To my horror, he slammed his face right into the pool steps. His cheek swelled like a balloon and he had an enormous black eye. With all that was going on, it was my proverbial last straw. Burying my head in the pillow, I cried all night.

The next day I was certain all the patients sitting alongside us in the dreaded waiting room were staring. They must have assumed Travis was the one who needed to be seen. His contusion was so bad, I really felt like a horrible parent for "allowing" it to happen — #momguilt.

Five long hours passed before Augi's name was called, but it was worth the wait to see "the wizard."

"I've never seen a child this young with uveitis," she told me.

Great! Rare for the rare. Just my luck.

"Dilate him at night," she said after the two-hour visit.

What a brilliant idea, but I felt dumbfounded and betrayed that no one had suggested this before. Trying to emphasize the positive, I focused on the fact that at least Augi could now see during the day — a profound change in his life that would not have occurred had we not sought another opinion.

While the reward was great in this venture, there are some negatives you should be aware of, if you choose to do likewise. The traveling obviously takes time so you have to factor in missed work into your expenses along with the costs of lodging and gasoline. Also, be prepared to wait hours to see the doctor. Each time we returned to Bascom Palmer, our average wait time was about five hours. Yikes!

Even with the long wait times and additional expense, taking the extra step to find a specialty center proved to be another great decision. Just the fact that our son could see again during the daytime was enough to take the sting out of the guilt over some of our missteps. This not only helped Augi in his treatment protocol and his ability to see and read but would prove vital to getting through the dark days that were yet to come.

Chapter 10 – Get back on the horse

Once you have found the best pediatrician or general practitioner, a local eye doctor that treats rare conditions, and an eye center, you still can't rest on your laurels.

Uveitis has many causes, so a systemic disease is most often suspected. Patients who are diagnosed with "idiopathic" uveitis, well, simply said, just means the doctors just don't know the cause, but most likely there is or was one. Finding an underlying condition can be crucial to properly treating uveitis, and even more so to treating the body. I'll expand more about the causes/triggers later.

When your loved one suffers from a rare or mysterious illness, ordering medical records is extremely important. Doctors have so many patients, and often do not have time to carefully review these documents. In my experience, rarely have I seen a physician read a medical chart. While I'm sure some may read the charts prior to or after an appointment, I am convinced some simply opt not to read charts due to time constraints. This belief is based on doctors' comments and some of the absurd questions I've been asked about Augustine's illness. It became quickly apparent some of these doctors knew little of anything about my child's medical history. This is not an exaggeration and is rather alarming.

Even more concerning, some of the doctors have made notations about examinations that were never performed. Generally, there is a review of systems and one can only presume some of these are routine and perhaps noted as fine, even if not examined.

Anyway, once obtained, read the documents with an investigator's eye. Look for key terms or any unusual sign or symptom, and keep a list. One doctor may observe an anomaly and note such in the record for example, and a subsequent specialist may overlook it, rendering the notation useless. Such entry, if read, could, in fact, lead to a diagnosis or a greater understanding of your child's condition. So, it may be up to you to make a list of all pertinent notations to later present to your child's primary care physician.

Journaling is also extremely important. Experts say ninety percent of diagnoses come through the history. If your family traveled weeks or months prior to the onset of symptoms to South America for example, this would be an important piece of information to relay to an infectious disease doc. If your child had a rash or joint pains as a toddler, this also could provide a vital clue to elucidate an underlying cause such as juvenile arthritis.

It is also imperative to construct a detailed family medical history. If your aunt suffered from multiple sclerosis or Sjogren's syndrome, a rheumatological cause may be more suspected. If your family is of Mediterranean descent, perhaps Behcet's disease may be considered.

Think of yourself as a detective. Interview all your "witnesses" (relatives) in person if possible, and ask specific, probing questions. It is important to ask each person about their relatives as well. If you're interviewing your aunt, for example, ask her about your mother. Your aunt may remember something your mother does not, perhaps about a childhood affliction. This cross-referencing will ensure you are conducting a thorough probe.

Think of your child's mystery diagnosis as a suspect in a homicide case. You are the amateur sleuth attempting to unravel a cold case. Reading medical records, taking notes, journaling symptoms, and interviewing relatives is all part of the investigation.

If you come up with a few "suspects," try to exclude as many as possible. Further research on a condition may reveal your child was never in the geographical region to contract a certain infectious disease for example. Exclusion can be a great way to reduce the number of suspected conditions prior to approaching your doctor. Remember, you only have precious time with the doctor, so the more precise the discussion, the better. Plus, the fewer tests the doctor orders, the more you can reduce the costs associated with laboratory fees.

If you are in the unfortunate circumstance of having to search for a mystery diagnosis, there are some pitfalls to avoid. Some doctors and nurses experience a mild form of hypochondriasis during medical school, as they are exposed to so many signs, symptoms, and diseases. Try to keep this in the back of your mind when reading some of these horrible conditions, as it may cause some unnecessary worry.

Also, try to realize that if your child's doctor has failed to come up with a diagnosis, there is still hope. Medical books and references are no longer tucked away in some far-away library, literally, all the information is at your fingertips.

It's also important to remember that as parents, we are not doctors and only read information in a one-dimensional setting. Experts not only learn medicine in textbooks but practice their craft in a three-dimensional manner. So no matter how valiant your efforts, your abilities will naturally be limited.

Once you have a shortened list, share them with your child's pediatrician. A great doctor will welcome your input, particularly if your child is afflicted with an unknown illness. Most likely, your doctor will be able to quickly reduce the list even further and order an appropriate workup.

Better to not overwhelm your child's doctor with too many suspected conditions or schedule too many appointments. As difficult as it may be, try to remember your child's doctor most likely has a lot of patients, and no matter how conscientious or caring he or she is, their time is limited. Ironically, the more exceptional your child's doctor is, usually means their schedule is even more demanding. After an initial workup is unproductive, most pediatricians would be agreeable with ordering tests once or twice a year, especially if you can make a compelling case.

Another consideration is medical testing can become expensive and frequent exams or laboratory visits can wear on your child's psyche. It is a difficult balance — the benefit of a diagnosis versus the costs.

While it is difficult sometimes to weigh how assertive or aggressive you should be in seeking an accurate diagnosis, if you do your detective work, you can assist an open-minded doctor, and increase your chances of obtaining an accurate diagnosis. Identifying an underlying cause of your child's uveitis can modify the treatment modality significantly. In many cases, treating the systemic condition can actually lead to a remission of uveitis altogether. Obtaining a diagnosis can vastly improve your child's prognosis and is at least worth a reasonable effort.

Chapter 11 –Charley horse

So what causes uveitis? The list is long, and although it is impractical for me to enumerate them all, I will try to provide at least the highlights.

Rheumatological disorders are generally thought of first, particularly in regards to children, as the underlying cause of uveitis. **Juvenile Rheumatoid Arthritis** or JRA, sometimes referred to as JIA, is the one that most often comes to mind. Why children acquire juvenile arthritis is still a bit of a mystery, but experts believe a genetic component is likely. The main symptom, of course, is joint pain, but fatigue, fever, and other symptoms can occur. JIA is important to treat because this form of arthritis can be aggressive, and deform the joints and affect internal organs. **Still's disease** is a rare form of arthritis that can present in childhood and usually presents with a high fever and a salmon-colored rash. There is also an adult-onset form of Stills.

Behcet's disease is another condition associated with uveitis. A form of vasculitis, symptoms include oral ulcers, and the disease affects multiple systems including the heart and lungs. Behcet's is a good example of how family history can elicit a diagnosis as such disease is common in the Mediterranean population. The Oral ulcers also are a hallmark feature and if such is noted in a patient's history, along with uveitis, this information can assist a doctor in making an accurate diagnosis. Many of these disorders are diagnosed based on a clinical presentation rather than with a test per se, so having all the pertinent history can be essential in finding an answer.

Systemic lupus erythematosus is always high on the suspect list. This autoimmune disease typically presents with joint pains, fatigue, and has the hallmark feature of a malar or butterfly rash. Patients often have an elevated anti-nuclear antibody rate, although a positive result does not always indicate lupus.

Connective tissue diseases such as **scleroderma** also cause uveitis and chronic cough, Raynaud's phenomenon, and tightening of the skin are some common features. Raynaud's affects the hands and feet causing coldness, numbness, and discoloration, and is thought to be caused by oxygen-reducing vasospasms.

Immunological conditions such as **IGG4 disease** are also associated with uveitis. Jaundice, fatigue, weight loss, and swellings can occur, although this condition, like uveitis, can be asymptomatic. Often this condition is identified after elevations in the blood are found in a specific immunologic workup.

Sarcoid Uveitis is often suspected if there are certain types of lesions detected on the retina such as punched-out chorioretinal scars. Fever, fatigue, weight loss and lung involvement are some of the symptoms. Arthritis and dermatitis can also occur. Blau syndrome, named after the doctor that described it, is a specific form of pediatric sarcoidosis which is considered a hereditary condition.

Chorioretinal scars can also suggest specific eye conditions such as **birdshot chorioretinopathy**. The scars on the retina appear as birdshot would on a target. Such condition usually occurs in adults over forty and carries a poor prognosis.

Irritable bowel disease and **Crohn's disease** also have been linked to iritis. Some symptoms include chronic diarrhea, weight loss, fever, and abdominal tenderness. There is a genetic test available for the NOD2 gene which is considered a genetic marker for Crohn's as well as Blau syndrome.

TINU or tubulointerstitial nephritis and uveitis is a syndrome with symptoms including flank pain, malaise, and fever; in addition, of course, to the symptoms of the uveitis itself — photophobia, redness, eye pain, etc.

Other rare diseases such as Glycogen Storage Disease type Ia, while not typically thought of as causing uveitis, can trigger such due to the **hyperuricemia** that is part of the clinical picture. Hyperuricemic uveitis has been observed, and patients with gout should be screened for uveitis.

Patients with any condition affecting the joints including any form of arthritis whether rheumatoid or gouty, the spondyloarthropathies, connective tissue disease, and so on, should be screened for uveitis. Those diagnosed with sarcoidosis, irritable bowel disease or similar pulmonary or gastrointestinal disorders should be seen as well. Any person recently diagnosed with a condition that affects the kidneys such as nephritis should have their eyes examined also.

There are also several self-limiting pediatric and adult syndromes that are also associated with iritis/uveitis. **Steven Johnson's syndrome**, usually thought to develop as a severe allergic reaction, comes with a blistering rash and red-eye. **Kawasaki disease**, very similar in clinical presentation to polyarteritis nodosa, is usually a disease seen in young children, more often boys, in which the heart and eyes are affected. It is important for doctors and parents to distinguish Kawasaki Disease from Scarlet fever. The former has potentially life-threatening complications and the patient must be treated within ten days of infection to avoid permanent damage.

While there are many genetic diseases, syndromes, and rheumatological conditions that are known to cause uveitis, a careful review of systems and patient history should be useful in excluding most.

Chapter 12 – Zoonoses: As rare as rocking horse crap

In addition to the endless number of rheumatological conditions and genetic diseases that can cause uveitis, there is an even more daunting amount of infectious diseases that can trigger inflammation of the eyes.

If you are new to uveitis and medical research, one obstacle that you may find frustrating is that many of the conditions that cause inflammation of the eyes contain non-descript, similar symptoms such as joint pains, rash, fever, and red-eye. This applies to genetic conditions, syndromes, rheumatological conditions, and infectious diseases. Perhaps a brilliant epidemiologist one day will make use of these similarities and find a link, but until then, it muddies the water for those attempting to decipher the uveitis puzzle.

Whilst most of the infectious agents that cause uveitis are considered rare, this could be due to that fact that most of these conditions are under-diagnosed and often not reported; but irrespective of this, most are preventable. If your child has been diagnosed with uveitis and a rheumatological cause cannot be identified, a careful analysis of recent travel or exposure to certain animals could shed light on the culprit.

Geographical information alone could either increase suspicion of a certain condition or rule it out. For example, if you live in the northeastern section of the United States, Lyme disease should go on your "suspect" list. If your child hikes in the woods often, your suspicion should increase. A recollection of a tic bite and a rash should set off bells, making this condition your prime suspect. An important thing to keep in mind though is that a positive Lyme titer could be a false positive, and a negative one still does not completely rule out the condition. (This is a very important point to remember — in medicine, there are false positives and false negatives, so if there is a compelling enough reason, retesting or finding an alternative means of analysis may be necessary.)

In another circumstance, a child may have traveled with family to a country where certain infectious diseases thrive. Countries such as Colombia and Venezuela, for example, have a higher rate of Dengue fever. Also known as breakbone fever, this tropical disease causes high fever, rash, and joint pain. There have been patients who contracted uveitis even after recovering from this mosquito-borne disease. In fact, this most likely happens often because uveitis can be "quiet" and asymptomatic for some time.

While most people are aware of the infectious risks of travel, amazingly the risks associated with house pets are often not known. Dogs and cats can carry certain types of worms which can spread to humans and have devastating results. Children can inadvertently ingest the eggs of such, either through soil or even contaminated vegetables. Toxocariasis, particularly in an immune-compromised individual, can wreak havoc on the body causing abdominal pain, asthma, insomnia, and behavioral changes. These larvae can migrate to the eyes causing permanent damage, which can lead to blindness.

Most pregnant women have been warned not to change litter boxes, but children can also contract toxoplasmosis from contaminated soil or meat products. While most people with healthy immune systems are asymptomatic, even if infected; the very young and immunocompromised persons can become ill and experience eye complications.

Proper food preparation and good hygiene can prevent or minimize the risks associated with the aforementioned conditions.

Leptospirosis is another zoonotic condition that is transmitted to humans through horse urine. While most cases are mild, the bacteria, leptospira, can cause meningitis as well as produce ocular symptoms. This condition, also known as Weil's disease, carries a high mortality rate, even with treatment.

Histoplasmosis is a fungal infection that is generally contracted through the inhalation of soil particles contaminated by infected bird or bat droppings. Pulmonary involvement is typical, but in severe cases, the fungus disseminates throughout the body. Ocular involvement occurs along with all the other tragic sequelae such brings.

The raccoon roundworm, Baylisascaris procyonis, can be inadvertently ingested by children via contaminated soil. While extremely rare, this parasite causes **visceral larva migrans** in which the larvae travel via the bloodstream to various organs causing inflammatory changes and destruction. These insidious creatures can also travel to the eyes, a condition known as ocular larva migrans, leading to blindness.

Whenever someone posts a video showing humans feeding raccoons or allowing them into their homes, particularly with children around, I wince. Up to ninety percent of raccoons are definitive hosts of this nasty nematode, not to mention the risk of rabies. Prevention here is key, by not leaving easily accessible food items around and securing trash receptacles. Reinforcing good hygiene practices with children can also help minimize the risk of these types of infections. This applies particularly to small children who tend to play in the dirt and put objects in their mouths.

Certainly, your children will be exposed to various pets, perhaps horses, and sometimes even exotic animals. Obviously, it is impractical or even unthinkable to keep your children from the wonderful experiences animals bring, but avoidance of certain contaminants and good hygiene practices can greatly reduce any risks associated with such interactions.

Chapter 13 – I can make more generals, but horses cost money

There are many costs and downsides to labeling conditions as rare. As I mentioned earlier, doctors often think more within the "bell curve" and do not sometimes include the "rare" conditions in the differential. This leads to a lot of these ailments being under-diagnosed and therefore, underreported. It is safe to say the statistics on these conditions are inaccurate at best.

While I mentioned earlier the benefits of finding a specialty center are many, there are always the costs to consider. With great intentions, we took Augustine to Boston after the inflammation in his eyes continued for years. With no abatement in sight, we thought this was a wise choice. Unfortunately, he underwent a scleral depression which was rather traumatic, at least for me to watch, and the doctor later informed us that he would not treat our son long-distance. If you are planning to consult with a doctor who lives far away, consider asking this question in advance.

Often parents or patient advocates are instrumental in procuring a diagnosis for their loved ones, and if the information is not readily available or well-known, this can lead to a reduction in accurate diagnoses. Never have I seen a public service announcement on uveitis or other rare diseases for that matter. An education campaign aimed at doctors, teachers, veterinarians, and parents could increase the general awareness about how to recognize these conditions.

Some public service announcements also do not present the full story, so it is incumbent upon advocates to do their own research. Recently, there was a string of commercials paid for by an obstetric group aimed at warning mothers not to schedule their c-sections too early. What the experts failed to mention were the dangers of delivering a child late or past due-date, increasing the risk of meconium aspiration. Whilst it is true, it is risky to deliver a premature baby, it can be very harmful to a neonate to aspirate meconium, which can lead to cerebral palsy and even death. Parents should at least be given all the pertinent facts, so they can make an informed decision.

Sometimes doctors label conditions as "idiopathic," which means that a cause for such condition cannot be found. A rather costly and overused form of this is Sudden Infant Death Syndrome or SIDS. Working child crimes, I read quite a few autopsy reports involving children. When a pathologist labels an infant's death as SIDS, it really just signifies a cause could not be determined. It is safe to assume a lot of these babies die from sequelae of rare inborn errors of metabolism or other rare genetic diseases where central sleep apnea or cardiac events are common.

Enhanced metabolic screening for newborns is essential to reduce the number of these unexpected deaths. Furthermore, if pathologists were more aggressive in identifying these "rare" conditions perhaps more lives could be spared as siblings are often at a higher risk of death due to the genetic component of these illnesses.

Another unseen cost of labeling conditions as rare is the criminal element. There have been a plethora of cases in which parents have been improperly accused of child abuse due to an ignorance of rare disease. To illustrate, there is a case that highlights this point — a woman in Texas was falsely accused of poisoning her child with antifreeze. This mother was convicted of such and imprisoned as a result. As if by fate, she was pregnant at the time of her incarceration.

Born in prison, this inmate's child began exhibiting the same symptoms as the sibling who died months earlier. After some investigation, it was determined both children suffered from a "rare" inborn error of metabolism — propionic acidemia. Turns out the chemical makeup of the key ingredient in antifreeze and propionic acid differ only by one carbon atom. A laboratory error had cost this woman her freedom, and had it not been for the birth of another child, she may still have been imprisoned to this day.

There is a host of other conditions that can masquerade as child abuse. Children who suffer from congenital insensitivity to pain often injure themselves, casting suspicion on their parents. Osteogenesis imperfecta is a disease in which children have fragile bones resulting in frequent fractures. Ehlers-Danlos Syndrome causes cigarette paper scars to form and can resemble marks of abuse. Again, awareness and education amongst investigators are crucial here to avoid false allegations and arrests.

There are also many "rare" conditions that manifest with psychiatric symptoms, causing a person to be mislabeled as "crazy." Acute Intermittent Porphyria causes hallucinations and psychotic behavior, but such is not a psychiatric disease, rather a metabolic disturbance. Children suffering from PANDAS, pediatric autoimmune neuropsychiatric disorders associated with streptococcal infections, display obsessive-compulsive features and are often misdiagnosed with OCD. These kids are placed unnecessarily on psychotropic drugs and worse... are not treated properly for the underlying cause of their symptoms — strep.

Researchers have even postulated that mosquito-related viruses are responsible for schizophrenia. Indeed, there could be organic causes for some of the known psychiatric conditions not related to personality disorders caused by upbringing or trauma.

More psychiatrists could serve their patients better by being well-versed in the organic conditions that cause hallucinations, psychotic episodes, tics, etc. These professionals are, after all, medical doctors. It is unlikely, though, that many of them would think of AIP or PANDAS when treating a psychotic or compulsive patient. Again, this goes back to the need for improved training and education. While many outstanding psychiatrists and therapists have recognized these differentials, there is always room for progress.

With the ever-increasing rise in health premiums and medical costs, it is now, more than ever, critical to addressing this diagnostic crisis, if you will. The ascending expenditures associated with labeling too many conditions as "rare," and the high rate of misdiagnoses such thinking precipitates, is costing all of us.

Not only are the material fees a factor, but the personal costs are high as well. In fact, death from medical error is the third leading cause of death in the United States, exceeded only by heart disease and cancer. Simply put, doctors kill more people than criminals. Whilst this may sound harsh, it is necessary to understand the magnitude of this problem to fix it.

Concomitantly, there is a high emotional price to pay. Depression is a real and potential burden suffered by many who troubled with an unknown or mislabeled chronic illness. After many interviews with suicidal patients both in law enforcement and during my tenure at a psychiatric hospital, this is a common sentiment.

Pharmaceutical mistakes as well, whether improperly prescribed or filled, is another important consideration and a reminder for parents to be vigilant. One of my son's doctors inadvertently prescribed him a pesticide used to treat cancer by mistake, and luckily our pharmacist caught the error. The intended medication was a sedative for a surgical procedure, and the difference in the name of the medicine was only a few letters. This was one of many close calls.

Establishing a rapport with your local pharmacist and double-checking prescriptions is a way to reduce the risk of this type of error. Taking the extra step of photocopying a script is a good habit to get into, because once you turn in the paper, it's gone. Also, asking questions about the dosage and instructions can minimize common distribution errors that can occur at home. If you are unsure, you can also inspect the markings on the pills to see if they match the known or prescribed medication as an extra precaution.

The intelligence and dedication required to be a doctor cannot be overstated, however; this has caused doctors to be elevated in status by the public to almost god-like beings. Some assume doctors are infallible...almost omnipotent. This puts too much of a burden on doctors, and can inadvertently inflate their egos. This is dangerous because, arrogance, particularly when making life-and-death decisions, can lead to impulsive decisions that could prove lethal. This perception also can cause patients to blindly (forgive the play on words) accept any doctors' words or opinions as dogma, and this can prove fatal.

It takes a certain degree of humility to truly be an exceptional physician, the kind that is willing to include parents, nurses, and even other doctors in the evaluation process. Removing the "god-complex" associated with physicians could benefit the medical community as well by relieving some of the enormous pressure such a label brings.

Whether it is through research, double-checking prescriptions, seeking second opinions, and just asking questions, parents can advocate for their children and participate in the process to increase their child's chances of successfully navigating a chronic condition.

Changing attitudes towards rare disease, increasing education and training of medical personnel, and improving public awareness can drastically reduce the number of deaths, injuries, and costs associated with the under-diagnosis and improper treatment of such types of illnesses.

Advocates for patients can also reduce the costs associated with such types of conditions by their participation. The costs associated with caring for a child with a chronic condition are staggeringly high, so any methods to minimize such should be employed.

Chapter 14 – Back the wrong horse

The number of rheumatological, genetic, and infectious diseases that can trigger the inflammatory process in the eyes is intimidating. Red herrings are common in medicine and can lead you in the wrong direction. For example, a child with uveitis and arthritis may lead to an improper diagnosis of JIA, when in fact, the child's arthritis is actually caused by gout or infectious agents. There are actually so many diseases that cause joint inflammation itself, that arthritis could be secondary, further complicating the matter.

Red herrings can also pose a different diagnostic problem. Just as in homicide investigations, sometimes there are too many promising suspects that detract from the real culprit. In our case, Augustine had several conditions such as polyarticular arthritis, Steven Johnson's syndrome, exposure to certain infectious agents, etc. that could have caused his condition. This may be equivalent to searching for the Phantom at a masquerade party.

Thankfully, there are some clinical pearls in medicine that can assist in reaching an accurate diagnosis. In fact, some rare diseases can even be diagnosed by signs in the eyes themselves.

One of my dear friends was hospitalized at the age of seventeen with a life-threatening illness. The doctors were baffled, unable to discover the cause of her predicament. One of the docs even devastated her mother by breaking the bad news that her she had leukemia. Luckily, a family friend, who also happened to be a doctor, visited her and noticed something all the others had missed: the orange ring surrounding the iris of one of her eyes.

A Kayser-Fleisher ring is the perfect example of a clinical pearl. The dark ring that forms around the colored portion of the eye is indicative of Wilson's disease — a rare, genetic disease that affects the liver and the joints. Copper accumulates in the brain and is responsible for the deposits in the eye as well. While this condition is treatable, it is often missed. In my friend's case, if it had not been for this chance encounter, she most likely would not have survived. The eyes aren't just the windows to the soul, they can be a portal to diagnosis.

In gouty arthritis, inflammation most often occurs in the big toe, and if such area is affected, this can signal the condition. If a person awakes in the middle of the night with pain in their big digit, gout is most likely the culprit.

A strawberry tongue is a hallmark feature of Kawasaki syndrome. While the epidemiology of this condition is poorly understood, following the treatment protocol for such is vital to minimize damage to the coronary arteries. Kawasaki, like a host of other self-limiting syndromes, has dire effects on the body and this truly amplifies the need for an effective diagnosis. When not treated in a timely fashion, the consequences can be serious and permanent, if not fatal.

Worth noting, children can present with atypical presentations even of a rare disease, drastically reducing a proper diagnosis and worsening the prognosis. Case in point, my daughter suffered from Kawasaki at age eight, outside the typical age range and gender associated with the syndrome. Luckily, it was recognized by the ninth day, just inside the recommended treatment window. When I learned John Travolta's son had also suffered from Kawasaki, this was the only time outside of medical literature that I had heard of a case. Another lonely disease...just what the doctor ordered, not. It seems we have a lot of zebras in our family.

On a side note, Kawasaki is most prevalent in Japan, and the epidemiology of such condition is poorly understood. I've often wondered about the role of radiation or other environmental factors. Japan is, after all, the only country to ever be struck by two atomic weapons and it is completely surrounded by water. The United States is second in prevalence and Hawaii has the most reported cases. Interesting that the Pacific state is also surrounded by water but also has a significant Asian population.

Can the wind make you sick? LOL, but actually some experts believe the infectious agents that trigger Kawasaki is transported by the air stream. Either way, most epidemiological information is gathered by the survey, and most likely, not very accurate. On a serious note, though, this pediatric illness is the leading cause of acquired heart disease in children and more research into the cause is needed.

Rashes can be seen in various conditions that cause uveitis, and a proper description of the rash can assist in distinguishing the culprit. A salmon-colored rash is seen in Still's disease while a "sandpaper" rash is indicative of Kawasaki disease. A malar rash (butterfly distribution) is seen in Lupus. Rashes accompany many infectious diseases such as Lyme disease and generally follow a well-documented pattern.

While there are many red herrings in medicine that can lead to an improper diagnosis, thankfully there are also clinical pearls that can lead to an accurate one.

Chapter 15 – Beat a dead horse

If you have exhausted all the usual means of finding a cause for your child's uveitis, there may come a point in time to accept a trigger may never be identified. The key here is to still treat the uveitis aggressively and whilst this is frustrating, it does not necessarily mean your quest is over.

Just as in cold case investigations, sometimes the passage of time is actually a benefit. Some conditions cannot be identified until a certain progression occurs or until a symptom appears. Some diseases progress slowly and only non-descript, generic complaints emerge at the onset. Other conditions can remit and relapse, and a clear picture cannot be gleaned in the early stages. It can often take years for certain clinical pearls to crop up. Often doctors and patients will look back and the earlier signs make sense in retrospect.

Also, just as in cold case investigations, often a fresh set of eyes can determine a solution that baffled others. Doctors and parents alike can become so immersed in the minutia, a simple or obvious fact is overlooked. The KISS or "keep it simple stupid" philosophy can easily be forgotten when a case becomes complicated. It is also easier to analyze a medical mystery when there is not a deep emotional attachment.

The most important issue here is not to overly obsess about a diagnosis. If you've consulted with specialists, reviewed records, and made a serious effort, then it is time to let go. Beating the proverbial dead horse will only exhaust you and leave you less able to care for your chronically ill child. It's okay to let go.

Chapter 16 – Horse around

If your child has been diagnosed with uveitis, it is imperative to take extra precautions to avoid further injury. Protective eyewear is a must for outdoor activities, particularly contact sports. Rising pressures in the eyes are one of the common side effects of both the condition and the drops. Any blunt trauma to the face, particularly orbital fractures, can dramatically increase the pressure in the eyes, especially fragile ones.

Sunglasses should be worn at all times while outdoors as the disease and most notably the eye drops cause cloudiness of the corneas which can lead to cataracts. Often uveitic children are photophobic anyway, and even fluorescent lighting can cause discomfort.

Emergency eyewash is an essential item for your home and vehicle. Chemicals inadvertently splashed or sprayed in the eyes can further aggravate inflammation, and can in fact even cause it. It's best to familiarize yourself with the product ahead of time to insure the wash is dispensed accordingly. During a crisis, the mind tends to revert back to muscle memory, so practicing with the product or being familiar with such is prudent.

If your child wishes to participate in contact sports such as football or lacrosse, you may want to discuss this with your child's pediatrician and ophthalmologist. There may be some additional precautions or certain recommendations they may suggest.

Kids will be kids and "horse around" but if you are prepared, taking simple steps to avoid further injury will give you some piece of mind.

Chapter 17 – Lead a horse to water

One of the toughest decisions we faced was whether to put Augi on methotrexate therapy. Methotrexate is a form of chemotherapy and acts as an immunosuppressant. While I am a strong proponent of treating uveitis aggressively, in Augi's case, his immune system was already compromised from various other conditions in which he was afflicted. This drug would have posed a high risk of dangerous infections, even in a healthy person. Feeling strong side-effects would pose a serious threat to his life, we opted out of chemo, but this is just in Augi's case, and may not apply to your child or loved one.

This decision to not follow one of the doctor's recommendations terrified me, even though several of his other doctors concurred with my opinion. Because he was old enough to understand, my husband and I decided to include him in the decision-making process, but we still feared our choice could backfire. After all, if his vision continued to deteriorate and the inflammation became uncontrollable, all of our hard work would have been fruitless.

Augi was still taking NSAIDs and instilling the drops, but we wanted to do more to stack the deck in his favor so to speak. Since we opted out of the chemo, our focus shifted to some alternate forms of medicine.

Many experts have proposed over the years that diet and vitamins, even alone, can combat inflammation. Ancient cultures have recognized the benefits of herbs and certain foods in the battle against ailments and disease, so this is really nothing new.

There are many juicers and blenders on the market that optimize the nutrients found in produce. Also, there is a wide variety of books and websites that provide information on foods with anti-inflammatory properties. Organic fruits and vegetables are best of course, but remember to blanch them or wash these products thoroughly, as they carry a high content of E-coli.

Turmeric is indigenous to India and a member of the ginger family. It is widely accepted that this plant, often boiled and dried, suppresses inflammation. Used often in curry-style dishes, turmeric is also available in pill form.

Ginger, like its cousin, is often used in Asian cuisine and has been known to subside nausea. Taken on a regular basis either in food or in pill form, ginger can also reduce inflammation in the body.

Forget warding off vampires, garlic will scare inflammation too. Garlic is a practical addition to almost any meal whether Italian or not, and its benefits go beyond its anti-inflammatory properties.

The superfoods such as açai come in powder forms, and such are a nice addition to health-smoothies. Frozen berries, bananas, spinach, kale, and flax seed are also some great accompaniments.

Many nutritionists recommend a Mediterranean diet for conditions such as cancer and inflammatory disease. An important aspect to remember is such experts consider sugar a fuel to such destructive processes, so a regiment low in sugars, refined foods, and fats is recommended. Meals rich in grains, the healthy fats, such as Omega-3's, and colorful antioxidant-rich fruits and vegetables are very beneficial.

In addition to multi-vitamins, there are specialty vitamins and concoctions containing trace minerals created just for eye health. Vitamins C and E, Zinc, and beta-carotene promote healthy eyes amongst others.

Consulting with a nutritionist would be a great start in implementing dietary changes. The introduction of herbs and supplements is a great way to secure your child's recovery.

There are also many beneficial exercises, such as swimming, which will increase circulation while not putting too much strain on the joints. High-impact activities should be avoided if the joints are involved as localized as well as systemic inflammation can spread to the eyes. Yoga is also a great option for those afflicted with arthritis, but if you have a male child, well, this may be a hard sell. Foolishly, I tried...what a mom fail!

There are also other alternative forms of medicine such as acupuncture, massage therapy, and other holistic remedies. Meditation is also a great form of self-healing, and there are a plethora of apps available.

Psychoneuroimmunology, the study of the effects on the mind and how thought processes can affect the course of the disease, is a fascinating and up and coming field. Evidence suggests a patient's mindset can greatly impact one's health, so the placebo-effect can be put to good use.

Vitamins, supplements, exercise, a low-sugar diet, herbal remedies, and meditation used in combination can greatly reduce inflammation and the need to rely on dangerous drugs. However; the older your child gets, especially when he or she enters the teen years, this routine will become much more challenging to implement. Hence, you can lead your child to a smoothie, but it does not mean such will be consumed.

Chapter 18 – Talk off the hind leg of a horse

One of the biggest mistakes I've made over the years with this disease is venting too much to friends and family about it. Don't get me wrong...my friends are generous...the best, and I have a close-knit, supportive family, but the truth is, it's really difficult for those on the outside to truly appreciate the struggles brought on by this affliction. It is like belonging to a club, but instead of getting a spot to park your yacht or a nice plate of lobster, you get heartache and anxiety.

Simply put, it's a lonely society to be in, and even if you can find another member, chances are they live far away, in another state, maybe even in another country...but it's worth it to find them.

After many failed attempts of obtaining validation from outside the home, I learned that most just were not in the position to give me what I needed. Not only do people not want to talk about bad things such as divorce or handicaps, there are other barriers to finding support. Unlike other diseases with obvious afflictions that can be seen, uveitis is a rather silent disease. Augi looked normal, just like every other kid. The perception was, of course, that he was normal because he looked fine. So, naturally, when I spoke about the matter, often the reaction was rather ambivalent.

After much wasted time feeling slighted, I learned not to bring up that ugly word to family or friends. It was not that they did not care about me or my problems, but the degree of my suffering was just not realized. Fortunately, I have a very supportive spouse who has always been willing to listen to my concerns or ideas.

Whether you're single or find yourself in the same position that I was in, and have no one to vent to, there are a variety of online support groups and even some local ones too. Finding at least one person that's in your shoes, someone that can relate, can really help ease your mind and provide validation. Even if you can't find a uveitis support group in your town, there is most likely an Arthritis Foundation chapter that could be a good place to start.

Not only can people in a similar situation offer emotional support, sometimes they have already traversed much of the system and can offer great advice. You never know — someone you meet could even say just the right thing to shed light on an underlying condition you may be searching for or recommend a specific treatment.

In my son's case, I learned that switching brands of steroid eye drops (from Pred Forte to Durezol) reduced the inflammation without raising the pressures as much — something one can only learn through experience. If I could have learned this from another parent, perhaps his condition would have improved sooner.

Bottom line, if you're going to talk to someone regularly about your child's condition, try to find those who are in similar situations. Such persons will have a stronger emotional investment in the subject matter, and can hopefully offer much-needed advice and support. Finding an empathetic ear can make your journey much more bearable.

Chapter 19 – Put a horse out to pasture

Not only can support groups or systems offer much needed emotional assistance, advice, and such, it is really important as a caretaker to think of yourself. At times, I was so overwhelmed, my own needs were put aside. Schedule a massage, take a bubble bath, or whatever you can think of to put the worries out of your mind. Your child or loved one will depend on upon you for years to come, and you won't do him or her any good if you're a basket-case.

Spending so many nights worrying or crying if an appointment went bad, was futile. Looking back, I wasted a lot of time. A control-freak by nature, the one thing I finally learned is even if you are a really involved parent: take notes, journal, find the best doctor, follow treatment protocols, etc. — there comes a point in time where you have to accept a harsh reality — you're not in control.

With any difficult diagnosis, one initially goes through the five stages of mourning. Lingering on anger the most, I expended too much energy blaming doctors for — missing the uveitis, for not disclosing all the conditions my son had, for dilating his eyes too long. Finally, though, I had an epiphany. It was time to let it all go. Even though it was overdue, I finally reached that final stage of acceptance. Realizing my child's fate was in someone else's hands was vital to moving on, and forgiveness was an integral part of the process. Freeing myself from all the bad feelings was empowering.

Sometimes it helps to turn control over to a higher power. Whether this is faith in God, the universe, or whatever your source of spiritual comfort, relieving yourself of command is imperative because it was never yours. Whatever your belief system, it really helps to let go. Do all you can…but accept it's out of your hands. Prayer helps or perhaps meditation…even yoga…if that's your thing. Find a spiritual outlet that can help you accept the uncertainty of it all.

If your anxiety is too much to cope with, don't be ashamed to seek help. People seek therapy for a lot less.

Chapter 20 – For one to fly, one only needs to take the reins

If you are caring for a child with a chronic condition such as uveitis, chances are some unexpected learning complications will arise, as well as some unanticipated red-tape. Being prepared will help to cut through most of it.

Children with eye disease often must take eye drops several times a day, often during school hours. Not only is it important to have your child's doctor complete the necessary medical authorization forms, it is vital to schedule an in-person meeting with the school nurse. The administration of drops is not always performed properly, and such is crucial to ensure your child receives the medication as directed. If done improperly, the medication loses much of its effect. It actually surprised me there is an actual right and wrong way to put drops in the eyes. Who knew? Most ophthalmologists carry pamphlets with diagrams and instructions.

If you leave your children with caregivers, even relatives, it is important to draft a consent form allowing any guardian to seek medical care for your child. Some hospitals and doctors will not treat minors without such forms, and in the event of an emergency, any delay could be catastrophic. Better to have the form notarized to avoid any issues.

The Rehabilitation Act of 1973 is a federal law that prohibits discrimination against people with disabilities. Any program that receives federal funds must adhere to the guidelines established in this law. Section 504 requires employers and schools to make reasonable accommodations to employees or students with disabilities. Most school administrators refer to this as a 504 plan.

Some students can receive extra time for test taking or completing homework due to low vision. Others can have reading materials or tests in larger fonts if necessary. Regardless of the degree of severity of the disability or accommodations needed, it is important to meet with your child's guidance counselor to discuss any special needs your child may have.

If your employer offers a healthcare savings plan, I would highly recommend signing up. The average expenses for caring for a child with uveitis is at least $100,000 and could be more depending on the severity of the condition. This estimate is just for childhood, and if there are other systemic problems, there will be additional expenses.

There are also many federal and state financial assistance programs available to children with disabilities. While I did not personally utilize any such programs, it was nice knowing they were available if the need would have arisen.

Once your child reaches high school and college age, there are many grants and scholarships available to children with uveitis and arthritis. Caring for a child with a chronic condition is extremely expensive, so finding programs such as these could allow you to recover some of your costs.

Another aspect to consider is to monitor your child's psyche. Augustine had to wear sunglasses during class during his middle school years. This was due to his extreme photophobia at the time. Some of the kids teased him and called him Stevie Wonder and Ray

Charles. Augi had a very good sense of humor, and also remarked he thought it was an honor to be referred to as these musicians he so admired. Well, the point here is that he most certainly must have felt "different" at times.

The multiple eye appointments, extensive waiting times, surgeries, procedures, traveling, and trips to specialists can become rather grueling. It is extremely important to monitor your child's psychological health and consider counseling if you feel appropriate. Make sure your child feels comfortable expressing any feelings he or she may have regarding their condition and the challenges such brings.

Chapter 21 – Horse laugh

Finding a healthy balance in caring for a child with uveitis can be difficult to achieve. Are you doing too much? Not enough? Just like anything in life, moderation is key.

Managing your child's condition can be stressful and not allowing this illness to take over every aspect of your life can be quite a challenge. Once you have found the best pediatrician, a local doctor that is experienced in rare disease, and an eye center for consults and emergencies, there is a practical approach to ensure both you and your child's needs are being met.

If you've consulted a rheumatologist and an infectious disease specialist and you're still in the dark as to the cause of the uveitis, there is still hope. As I mentioned earlier, keeping a thorough journal, and conducting some research could eventually bring the answers to light. Be cautious about reading too much about uveitis itself as this may only frighten you, and does not serve much of a purpose. Only an eye doctor with their specialized equipment can care for your child's eyes, so best to leave this to the experts.

On the search for a cause, though, cost and your child's psyche should be factored into the equation. While it is true, finding a cause could drastically improve the outcome, the reality is such a quest can be expensive, time-consuming, and wear on your child...even the entire family.

The eye center we selected is five hours away, so we tried to turn these trips into mini-vacations for the kids as best as we could. There are certain hotel chains that provide free breakfast. Our kids loved the buffets and it became a running joke to see who could stack their plate the highest. Whoever knocked on the hotel room door would yell out "housekeeping," in a dainty female voice.

Laughter is a great remedy so finding inventive ways to pass the time will make the wait less mundane. Creating a jovial and carefree atmosphere reduces anxiety and stress. A sense of humor goes a long way in these situations and helps to lighten the mood.

Wait times for doctors can be lengthy and this is even truer at specialty centers. For us, the wait to see the retinal expert was a punishing five hours each time. Planning ahead by bringing handheld games, snacks, drinks, and such, made these long waits much more bearable.

Chapter 22 – Horse and buggy

Surprising that with all of the modern advances in technology and medicine, no one has yet created a computer program that can sift through and diagnose some of these rare diseases. Let's face it...there is an app for everything. Want a ride somewhere...click on Uber; hungry...there are drivers that will bring almost any gourmet offering to your door. So why not an app to diagnose your condition? Not only is it logical, necessary, and cost-effective...I think it's coming.

While the diagnosing of the disease may be too complicated for an app, certainly, software could be developed to more expeditiously go through the thousands upon thousands of conditions. An algorithm could be formulated with "if and then" statements that could be fairly accurate. Conditional constructs could be formulated in such a way as to accelerate the diagnosis process. It could be a very useful tool for busy doctors and for patients desperate for answers.

Most people first seek medical advice when they are not sure what is wrong. Symptoms are often non-descript, so it is not uncommon for doctors to order tests to assist in the decipherment process. It is no wonder medical costs are so staggeringly high. Realistically, often the doctors are as in the dark as much as the patients, particularly if the problem is complex. A computer processor could sort through symptoms and search thousands of conditions in seconds, providing possible culprits which would reduce unnecessary laboratory procedures and expensive tests.

Another way that computers and apps could greatly modernize medicine, is to store medical records in a central location. It can take months to obtain, organize, and scan all one's medical records. Wouldn't it be great if they were all in one place? There are countless scenarios also in which a person may be unable to communicate. If doctors had access to one's records irrespective of location, emergency information such as contacts, allergies, medical history and so on, could be available.

Doctors will always be needed to order tests, evaluate results, perform surgeries, etc. It would be beneficial to patients and doctors alike to at least simplify the diagnosis process, and what better way than a computer program that allows for the input of all signs, symptoms, diseases, and the like.

There is too much information available for any one doctor to possess all of this knowledge, so this is a must for the future. Accurate and efficient diagnoses will reduce medical costs and save lives. A new paradigm for diagnosing patients and storing medical records is needed to save money and increase efficiency, which will modernize the medical field to keep pace with the technology in our millennium.

Chapter 23 – Lock the barn door after the horse has bolted

With emerging diseases such as Zika, it is imperative to spread the word about the dangers of uveitis and the infectious diseases that cause it. Whilst I have yet to see any mention of Zika causing Uveitis, the tropical disease does affect the joints, creating a high index of suspicion. In fact, during the course of editing this book, I discovered my suspicions have been confirmed — Zika has been linked to uveitis. This tropical infection is spreading to North America, so the timing of this publication is rather eerie. While there has been much media coverage regarding the risk to unborn children whose mothers are infected with Zika, there has been no mention of the risk to older children's vision. This is completely unacceptable.

Zika transmitted via the placenta to the fetus can cause microcephaly in the neonate. Tragically, blindness can occur in microcephalic children. Ironically, Zika causes blindness in two ways via microcephaly and uveitis, so efforts at stopping the spread of this disease are crucial.

Zika has also been associated with the development of Guillain-Barré syndrome, an autoimmune reaction that affects the nervous system. Paralysis, weakness, and pain can occur along with respiratory failure. Those with autoimmune disease or tendencies may be at a higher risk for acquiring this complication.

Zika, along with other mosquito-borne diseases, is preventable in many cases. Removing items with standing water from around your home, wearing long-sleeves and pants while outside, using mosquito repellent products containing DEET or other approved chemicals, and avoiding being outdoors during certain hours are some means of preventing mosquito bites. Living in a swamp, I am unable to remove standing water, LOL.

Being able to recognize the type of mosquito known to spread Zika, Aedes Aegypti, is important. This mosquito is larger than the typical mosquito and is black with white markings.

The Centers for Disease Control has posted recommendations on their website — www.cdc.gov. It is imperative to check such site regularly for updates, particularly for those who reside in one of the affected regions.

With the increase in travel and globalization, even the older, well-documented infectious diseases pose a risk. Re-emerging diseases such as measles will also increase the number of uveitis cases, therefore, increasing the rate of blindness. With the recent release of the documentary *Vaxxed*, there could be an even greater threat of more blindness-causing disease due to an increase in the number of parents not inoculating their children.

While it is certainly not my place to judge how others raise their children, I would caution any parent considering opting out of vaccinations to at least review the consequences of such diseases and the horrific suffering these illnesses cause. Sympathetic, I was also nervous about potential side-effects vaccines can cause but death, disfigurement, and blindness are potential sequelae of childhood diseases such as measles.

As if the spread of infectious disease were not enough to raise concern, the ever-increasing threat of terrorism adds to the need to be vigilant. Biological agents and chemical warfare are real threats facing all of us. Many infectious agents have been weaponized and some are associated with uveitis. Tularemia, an infectious disease contracted by ingestion of contaminated soil or contact with an infected animal, has been weaponized. Anthrax, another bacterium, has also been used as a biological warfare weapon.

Emerging diseases and those re-emerging ones due to a decrease in vaccinations may significantly increase the number of uveitis cases, increasing the rate of blindness worldwide. It is incumbent upon government officials, doctors, and legislators to address these concerns.

Chapter 24 – The horse's mouth

While horses and other animals can carry viruses or bacteria that transmit the disease to humans, these mammals can also contract such infections and uveitis as well.

Moon blindness, or equine uveitis, is the most common cause of blindness in horses. The zoonotic infections that threaten humans exposed to horses or mosquitoes are likely involved in the acquired uveitic cases in horses such as leptospirosis or West Nile virus.

Dogs and cats can also acquire uveitis, along with, I presume, most pets.

Veterinarians could play a larger role in the education of pet owners regarding not only the threat to their animals but to the members of the household as well. These medical doctors are best suited to warn parents about the potential risks associated with pet ownership and the common sense preventative measures that can be taken to avoid zoonotic infections.

Veterinarian associations could also participate by distributing educational material to animal clinics to warn parents about zoonoses that cause blindness and sickness.

In fact, the Centers for Disease Control could do more in educating the public about such diseases either through television, social media campaigns, or other means.

Chapter 25 – On one's high horse

While it is imperative to improve education, awareness, and screening, it is even more critical to increase funding for low-income children and those in developing countries. It is astounding in our modern era, with all our sophisticated technology and resources, how many children go without basic care.

Even in an ideal setting, these eye conditions still go undetected and untreated. So, one can only imagine how dire the circumstances are for those in the developing world.

Walgreens recently launched a program in which they are donating a portion of their proceeds to Vitamin Angels, a charitable organization whose mission is to provide vitamins and nutrients to children in underdeveloped areas. Vitamin A deficiency, while very rare in most countries, can cause blindness. Just this week, I was made aware of this media campaign aimed at preventing blindness in children...what a great cause!

Hopefully, by increasing public awareness about these types of conditions, we can prevent blindness not just locally, but globally. The more people who learn about the devastating effects of blindness brought on by malnutrition and exposure to infectious agents, the better. Hopefully, this cause will awaken many generous spirits.

Chapter 26– Horse trading

As I've mentioned previously, I have never seen a public service announcement about uveitis. This is not to say there has never been one, but if there has, they have been scarce.

With uveitis being the leading cause of blindness in children this is nothing short of shocking. Public awareness is needed because most of the conditions that cause uveitis can be prevented. Also, the recognition and early treatment of uveitis is critical in procuring a timely remission.

Blindness has to be one of the most debilitating disabilities, not to mention one of the most frightening. Prisoners of war often report being locked in dark rooms as a form of torture. Often these brave heroes state this to be the most terrifying of any torment. As horrific as this sounds, for those deprived of sight, there is never a return to the light.

Perhaps with increased funding and advances in research, there may be hope for blind persons. There have been tremendous advances recently in restoring hearing to deaf persons, so likely the technology will advance enough to restore vision to the blind.

Chapter 27 – Dark horse

The casket was perched atop the bier. The funeral director removed the mort-cloth and invited the mourners to view the decedent. The bubble had burst — my beloved father was dead.

My father had the life-force of a hundred men, and his passing would leave a void that could never be filled. But as fate would dictate, even my mourning would be interrupted by that dreadful disease.

Even though I was hesitant to see him, some unseen force was beckoning me closer. As I walked through the crowd towards him, my husband, Quinn, grabbed my arm.

"It's bad. The doctor wants to see him immediately."

Augustine had woken up earlier that morning with an intense headache. Seeing flashing lights and little else, we panicked and called his ophthalmologist.

My husband took my son and left. A profound feeling of emptiness consumed my soul. Kneeling down at my father's side, I begged him to intervene. Now that he was dead, somehow, it did not seem that farfetched that he could.

The boys returned about an hour later with the somber news — emergency surgery.

The drive to Miami was long and excruciating. My eleven-year-old child could barely see, and I could think of little else. The future seemed bleak.

We stopped at a local "mercado" near our hotel. *American Pie,* one of my dad's favorite songs, was playing throughout the store. It was comforting to feel his presence, even if through music.

While awaiting an eye ultrasound, I walked over to the nurse's station to turn in some paperwork. Clueless that Augi had followed me, I turned around and inadvertently poked him in the eye. My efforts at controlling the floodgates were ineffectual. The receptionist that brought me into her office to complete the financial forms must have thought I had lost the plot.

During pre-surgery procedures, one of the nurses instilled drops in the wrong eye, apparently confused over patient's left or her left. Thankfully, my husband and I were there to catch the mistake, but this just demonstrates another important point. Even amongst trained professionals, mistakes can happen, so it is very important to double-check medications and such.

Augustine was quite brave as he was wheeled back beyond the steel surgical doors. When he awoke from the anesthetic, the tears rolled down his cheek, and it was one of the worst moments of my life.

During our stay in Miami, a concerned friend called me with some troubling news. Little did I know that whilst I was away from work, a jealous co-worker was bad mouthing me to anyone who would listen, including my captain. The manipulator's attempts were unsuccessful, but the cause of much-unneeded grief, but that's a story for another book.

Thankfully, though, the operation was a success and one of Augi's ophthalmologists commented years later on the surgeon's handiwork. This comment reinforced my opinion that there is a difference between a good surgeon and an outstanding one. Personally, the one who implanted my son's lenses seemed like an angel to me and is one of the only persons I would ever entrust with my child's eyes. So even through our darkness times, the light somehow shone through.

Chapter 28 – One's high horse

C.S. Lewis eloquently said, "I have seen great beauty of spirit in some who were great sufferers. I have seen men, for the most part, grown better, not worse with advancing years, and I have seen the last illness produce treasures of fortitude and meekness from most unpromising subjects."

Not that I adhere to Gnosticism or Stoic concepts, but I have always been fascinated by the philosophy that all is somehow "meant to be." Perhaps this philosophical principle helps me to cope with forces beyond my control. During moments of solitude or reflection, I often wondered if somehow my son's suffering was "for a reason," perhaps to benefit someone else.

There is no doubt that my child has been transformed over the years due to his illness, and even though it's hard to admit, I think, for the most part, it has been for the better. His outlook is quite positive, and I suppose he has a greater appreciation for not just life, but his sight.

For me, Augustine's illness was a humbling experience. It taught me that no matter how high you are on life at any moment, something unexpected can creep into your world, and shake you at your foundation. The years of battling the "silent disease" of uveitis brought me a heightened sense of empathy for others, especially when it can be so challenging to see other's suffering.

As a detective, this renewed compassion reminded me to make every effort to treat others with kindness and dignity, regardless of their circumstances. Truly, I believe my struggles with this disease shaped me to be a better civil servant, increasing my concern towards others.

This is particularly true with my passion for investigating cold cases. Even though I had never experienced their pain, I could relate to the isolation brought about by such a tragedy. Whilst I had to learn how to traverse the medical network alone, these forgotten victims had to navigate through the criminal justice system on behalf of their loved ones. Often the family members of slain victims will initiate their own investigation, feeling frustrated over a lack of answers. So, there are many parallels between fighting for a loved one with a mystery illness and seeking resolution on behalf of a cold case victim.

Chapter 29 – Better get back on my horse

There is at this time no cure for uveitis, but it can be controlled and managed. There are no guarantees, even in my own son's case, because uveitis is a remitting and relapsing condition.

The more involved you are, though, and the more you confront this condition through expert advice, medical treatment, nutrition, and so forth — the more you augment your child's chances of obtaining remission and staying there.

It is my profound hope that you or your child can benefit from some of my experiences. If you have any questions or need to commiserate, please feel free to contact me at coldcaseconsults@aol.com. I'd love to hear from you!

Appendix A
List of important reminders

- Find a pediatrician/general practitioner that is experienced in rare disease;
- Find a local ophthalmologist that has experience treating rare conditions;
- Be on the lookout for warning signs of eye disease such as blinking, squinting, eye rubbing, headaches, droopy eyelids, red eyes, eye discharge, leukocoria (white orb), pupillary changes (including shape, size, and reaction to light), difficulty in reading, blurriness, sitting too close to the television set, etc.
- Take your child to an eye doctor, not an optometrist, by age four;
- Take your child immediately to an ophthalmologist immediately if signs emerge;
- Stock your home and car with an emergency eye wash kit;
- Know how to treat eye emergencies such as blunt force trauma, chemical spills, and impalement;
- Remind your child to wear sunglasses and protective eyewear;
- Keep a journal of all doctor visits, illnesses, and injuries;
- Photograph any visible signs, especially rashes or bites;
- Teach your child good hand-washing skills, particularly around animals;
- Prepare all food according to package directions, thoroughly wash produce, and blanche organic products;

- Set up a daily vitamin routine including those vitamins that support healthy eye function;
- Maintain a healthy diet for your child including the healthy fats, colorful fruits, and vegetables, and anti-inflammatory herbs and spices;
- Talk to your veterinarian and pediatrician regarding ways to minimize zoonotic infections contracted from pets;
- De-worm your pets, clean their bedding regularly, and give them frequent baths;
- Avoid areas with disturbed soil such as horse arenas with dirt, recently excavated areas, and farms that utilize heavy equipment.
- Frequent the CDC website for updates on emergency illnesses and ways to avoid infection.

Appendix B

Questionnaire

- Age of onset of uveitis/red-eye; (Can differentiate pediatric causes from adult-onset, such as Kawasaki Disease vs. diet related — Gout for example)
- Geographic location; (Very useful for ruling infectious causes in or out)
- Family history of rheumatic conditions;
- Patient history;
- Other symptoms that accompanied the onset of uveitis; (could indicate viral or infectious, even a syndrome)
- Recent rash;
- Recent insect bites;
- Recent travel;
- Exposure to pets/animals including dogs, particularly puppies, cats, chickens, horses, raccoons, rabbits;
- Recent exposure to exotic pets, zoos, or farms;
- Exposure to contaminated soil;
- Exposure to animal droppings including bats, chickens, and raccoons;
- Gastrointestinal symptoms?
- Review of systems?
- Any clinical pearls?
- Rule out red herrings.

Appendix C

A modest list of Clinical Pearls

Colors:

Bluish tinge to sclera –
Osteogenesis Imperfecta
Bluish tinge to skin – Methemoglobinemia
Greenish hue to skin – Hyprochromic anemia
Orange hue to skin – Hemochromatosis
Purplish hue to urine after exposure to
sunlight - Acute Intermittent Porphyria
Salmon-colored rash – Still's Disease
Tea-colored or dark urine – Myoglobinuria

Eyes:

Blue sclera – osteogenesis imperfecta
Chorioretinal lesions – sarcoidosis
Kayser-Fleisher ring – Wilson's disease
Keratic precipitates – (small, round, and
white) – non-granulomatous uveitis
Keratic precipitates – (large, yellowish,
mutton-fat) - granulomatous

Food cravings/aversions:

Aversion to sweets/fruits – Hereditary
Fructose intolerance Cravings of dextrose-based
candies - Hereditary Fructose Intolerance

Food cravings/aversions con't:

fructose intolerance
Ice cravings – Anemia, diabetes Insipidus
Intense thirst/excessive water drinking –
Diabetes Insipidus
Postprandial somnolence/greasy foods –
cholecystitis/pancreatitis
Salt cravings – Addison's disease
Seizures before breakfast –
Glycogen storage Disease

Heritage:

Ashkenazi Jews –
Glycogen Storage Disease
Mediterranean –
Behcet's disease

Rheumatological:

Arachnodactyly (long, slender fingers) –
Marfan's Syndrome
Hyperextensibility – Ehlers-Danlos Syndrome
Malar rash – Lupus
Oral ulcers – Behcet's disease
Pain/swelling in big toe – Gout
Sticky joints – Spondyloarthropathy
Tightening skin – Morphea Scleroderma

Psychiatric manifestations:

Bipolar affective/schizoaffective disorders – Velo-cardio-facial syndrome

Obsessive-compulsive behavior and/or tics– **PANDAS**
Psychosis/hallucinations -
Acute Intermittent Porphyria

Scents:

Fishy odor - Trimethylaminuria
Maple syrup/sweet smell in urine –
Maple syrup urine disease
Skin tastes like salt – Cystic Fibrosis
Sweet smelling breath –
Lactic Acidosis and Diabetic ketoacidosis

Signs/Symptoms:

"Bleeding skin" appearance –
Steven Johnson's syndrome
Hyperuricemia – Glycogen Storage Disease, type 1a
Right shoulder pain – cholecystitis/IGG4 disease
Posterior helical indentations/pits – Beckwith-Wiedemann syndrome
Sleeping for long periods of time/teenagers –
Kleine-Levin Syndrome
Spider nevus – liver disease
Stridor – Multiple Carboxylase Deficiency
Strawberry tongue – Kawasaki Disease
Tachycardia – POTS,(Postural Orthostatic Tachycardia Syndrome)

Kimberlymcgath.com
Zodiacsettlingthescore.com
Twitter: @kimmcgath
Facebook.com/mcgathproductions
Facebook.com/kimberlymcgath
Facebook.com/zodiacsettlingthescore
Instagram.com/kimberlymcgath
Pinterest.com/kimberlymcgath
Linked In: Kimberly McGath
http://bit.ly/1NL8inY

www.ingramcontent.com/pod-product-compliance
Lightning Source LLC
Chambersburg PA
CBHW022110170526
45157CB00004B/1564